ROBOTICS

STUART AND DONNA PALTROWITZ

JULIAN MESSNER · NEW YORK

JEM

A JEM BOOK

Published by Julian Messner, a division of
Simon & Schuster, Inc.
Simon & Schuster Building,
1230 Avenue of the Americas,
New York, New York 10020.
JULIAN MESSNER and colophon are trademarks of
Simon & Schuster, Inc.

Manufactured in the United States of America

Design by John Treworgy
Edited by Jane Hyman, M.Ed.

Library of Congress Cataloging in Publication Data

Paltrowitz, Stuart.
Robotics.

"A Jem book."
Bibliography
Summary: Describes various types of robots, their
purpose, and how they function. Also discusses
briefly the future of robots and some of the tasks
they will perform.
1. Robots—Juvenile literature. [1. Robots]
I. Paltrowitz, Donna. II. Title.
TJ211.P33 1983 629.8′92 83-13108
ISBN 0-671-44077-2
ISBN 0-671-49720-0 Pbk.

To Adam Jay and Darren Scott
who never lack feedback

ACKNOWLEDGMENTS

We wish to express our thanks and appreciation to the following companies and individuals for their assistance in providing photographs and resource material:

Gene Beley, Android Amusement Corporation
Monrovia, California

Dr. Michael Freeman for his pictures of 2-X-L

Quasar Industries for pictures of Klatu

Cincinnati Milacron for T^3 photographs

Robot Institute of America

Joseph Engelberger of Unimation, Inc.
Danbury, Connecticut

Chrysler Corporation

National Aeronautics and Space Administration

The Museum of Modern Art Film Stills Archive

Photo Division of the CBS Television Network

CONTENTS

FOREWORD

You may not know it. There is a robot in your house. Don't be frightened. Look in your car. Do you see a robot? There's one in the supermarket. Robots are helping to make life better.

Robots aren't dangerous. They don't make scary sounds. Robots aren't alive. They don't have evil brains or kill their masters. Those kinds of robots only exist in the world of make-believe.

A robot is a mechanical object. It can perform human tasks. While doing tasks robots may not be directly controlled by a human. They follow orders. They can work without a human being there.

Some everyday things are robots. A wind-up toy is a robot. You wind it up and it moves. So is a mouse trap. You set it, and if a mouse touches the trap it snaps shut. Both are performing human tasks. The wind-up toy is moving without being carried. The mouse trap catches the mouse. A person doesn't have to wait for the mouse to appear. Of course, these are simple robots.

Not everyone likes robots. In fact, many believe they aren't needed. Some people think that humans should do their own work. The truth is that we all need robots. Many jobs are dangerous. Every time a job is done, there is the risk of injury. A worker could be hurt. Some jobs are also boring. People don't want to do the same task over and over for eight hours each day. A robot will do a dangerous job. It will also do a boring job. A robot doesn't have a mind of its own.

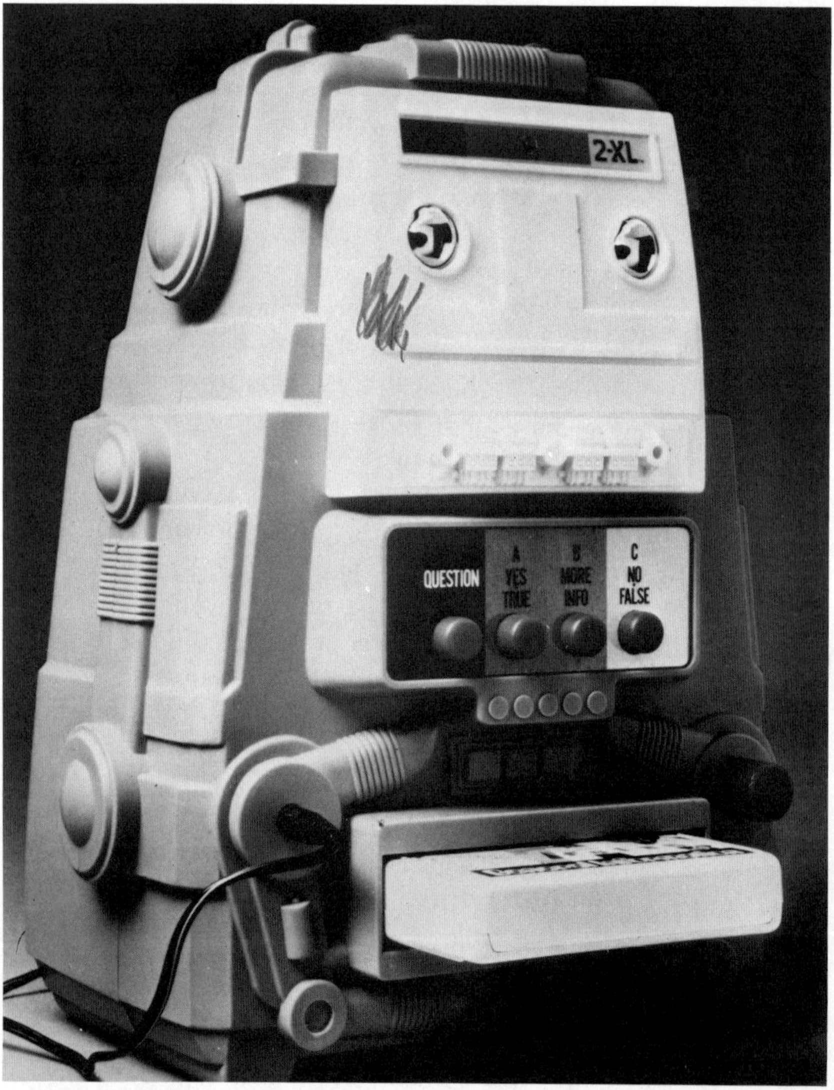
2-XL
QUESTION
A
YES
TRUE
B
MORE
INFO
C
NO
FALSE

Some people think "robot" is an old word. It isn't. The word "robot" is only about sixty years old. It came from a play in Czechoslovakia, which is a country in eastern Europe. *R.U.R.* was the name of the play. *R.U.R.* stood for Rossum's Universal Robots. Dr. Rossum had an invention. He made workers who were mechanical, but they looked human. These workers were called robots. Robot is a word in the Czech language. The word means "worker." It's a person forced to do work. In the play, the workers did as they were told. First they did all the world's work. Later they fought the wars. At last they rebelled by killing their masters.

Today robots aren't killers. They're helpers. Robots are making life easier. New uses are being found for robots. Almost anything done by a human may be done by a robot in the future.

The 2-X-L is a robot that was invented by a teacher to help students learn.

1
PERSONAL ROBOTS

ROBOTS THAT READ

If you're in the library and someone talks, don't say "shhh!" Look first. It may be a robot. There's a type of robot that reads books out loud. It's called a KRM or Kurzie, for short.

KRM is the Kurzweil Reading Machine. It looks like a photocopying machine. It doesn't look like a robot but it is one. The KRM was made by a computer scientist, Raymond Kurzweil.

This robot is special. It has more than 2000 rules about reading and saying words that are stored inside it. A person puts a book face down on the machine and then puts on headphones and listens. It can be made louder or softer and can be made to read faster or slower. Even the sound of the voice can be changed. There's nothing in the library that the KRM can't read.

The KRM is being used throughout the United States. It was used first in the Mid-Manhattan Library in New York. Now there are more than one hundred in use. The KRM was first made to help the blind read. Slow readers can be helped by the KRM, too. A KRM costs about $20,000. Yes, it can read any book. There is one thing this robot can't do. It can't write your book report.

ROBOTS THAT TEACH

Do you know that there's a robot that teaches? The robot is called 2-X-L. It has this name for a good reason. It will help you *to excel* in your work. This robot was invented by a teacher, Michael Freeman. The robot has an eight-track tape inside it. If you put a sports tape inside, the robot will talk about sports. You can put a science tape inside so that it will talk about science. Choose whatever subject you like.

Dr. Freeman made the robot to teach children. On the tape he put things that students needed to learn. He knew what they couldn't do well. On the robot's chest was a telephone dial. Each student had a number. The student dialed his or her own number and then put on headphones. Next it was time to listen to the robot. There were four buttons. The student pushed the buttons to make 2-X-L speak. The robot said "correct" for a right answer. It said "bad news" for a wrong answer. For more

Dr. Freeman, the inventor of the 2-X-L, is helping two students use his robot.

information there was a different button to press. The robot never hurt any feelings. It said, "I like you anyway."

The 2-X-L is only twelve inches high. It does a big job. Dr. Freeman thinks it will be great for schools everywhere. The 2-X-L is now made by the MEGO Corporation. Most kids play with toys. This toy plays back and teaches. More than 200,000 have been sold. Who knows? Maybe one day 2-X-L will be teaching you.

PET ROBOTS

If you'd like a pet, why not get a robot? A robot is easy to keep. Robots are small. They don't need any food. They do whatever you want them to do. Robots can be built right in your home. By using your imagination you can create your own. There are also robot kits people can buy.

Many people build their own robots. Robert Profeta is fifteen. He just built an amazing robot. It looks like R2D2. It can do many things. It tells Robert if water is clean to drink. It also tells him if someone is breaking into his house. The robot buzzes. It blinks, too. To build the robot Robert used a tin can, a dome, and some other parts. In all, it cost less than one hundred dollars.

Orion is the name of another computer pet. It was built by Terry and Ron Palmer. The robot can talk. It can ask questions. It can answer people in other languages. Orion asks a woman for a kiss. When he gets it, his antennae shoot up. He starts to spin around a few times. Not everyone has enough money for an Orion. Orion costs a lot. He cost $7000 to build.

Terrapin is a name that sounds funny. People stop laughing when they see it work. Terrapin looks like a turtle. It's a special robot. It can walk around the room. While it walks it measures the size of the room. Then it can draw a picture of the size of the room. Terrapin is only eight inches long. It's connected to a computer with a cord.

Terrrapin doesn't work by magic. It's hooked up to a small computer. The person with the computer tells Terrapin what to do. Terrapin is a robot kit. You can build it in your home. It costs about $400 to put together.

Then there is Boris Handroid. Boris plays chess. Boris knows where the other player moves. He has an arm to move his piece. How does Boris know? The chess pieces have magnets under them. The magnets tell sensors under the board where the chess piece was moved. Then Boris makes his move. Boris does not always win. Maybe for $1700 he should. Robots can lose, too. If he loses, Boris shakes hands. Boris is a gentleman.

Some people call them personal robots. Others call them smart pets. Whatever you call them, these robots are great. They do so many things well. It may not be long before we all have a personal robot. Watch out, R2D2.

This 2-X-L is teaching fifth graders at a Bronx, New York public school.

2

HOW ROBOTS WORK

Every kind of robot gives feedback. Feedback has nothing to do with food. It means giving information to the person or machine who controls it. Every person gives feedback. You eat a new food. You see a movie. You try on new shoes. You like it or you don't like it. When you tell someone your feelings, you're giving feedback.

You react to things. You drink hot tea. Then signals are sent to your brain. It's hot. You put the hot tea down. You reach for a cookie. You can't get it. Your hand tells your brain, "I didn't get it." Your brain says, "Reach higher for it." You reach and grab the cookie. The nerves in your hand gave feedback to your brain. Your brain controls your hand's behavior. Every time feedback is given, your brain tells your body what to do. Then your body acts.

The thermostat on the wall is a robot. It gives feedback. Inside the thermostat there's metal. When the room gets colder than you want it, the metal bends. As it bends, it closes an electric switch. The switch turns on the furnace or oil burner. When the room is heated to the setting on the thermostat, the metal bends back. The electric switch is opened. The oil burner or furnace goes off.

The automatic traffic signal is a robot, too. There's a pressure sensor. It's in a metal plate in the road. A car rides onto the metal. It gives feedback to the traffic control box. The box changes the light. The red light turns green.

Another robot is the electric fuse. It controls the electric power in a house. At the end of the fuse is metal. Sometimes the wires in the house get too hot. Then the metal on the fuse gets hotter and melts. Next the power shuts off. That way the wires don't catch on fire. The wires gave the feedback to the fuse that they were hot. The fuse took the feedback and shut off the electricity.

The thermostat, automatic traffic signal, and fuse are robots. They're robots for one reason. They have feedback. They're able to control themselves. To be able to give feedback they need two things. They need power. They also need parts that can move.

Working robots are all around us. Here's a list of some working robots. Can you think of more?

- automatic toaster
- automatic door opener
- smoke alarm
- central air conditioning
- burglar alarm
- automatic lawn sprinkler system
- automatic water heater
- automatic record player

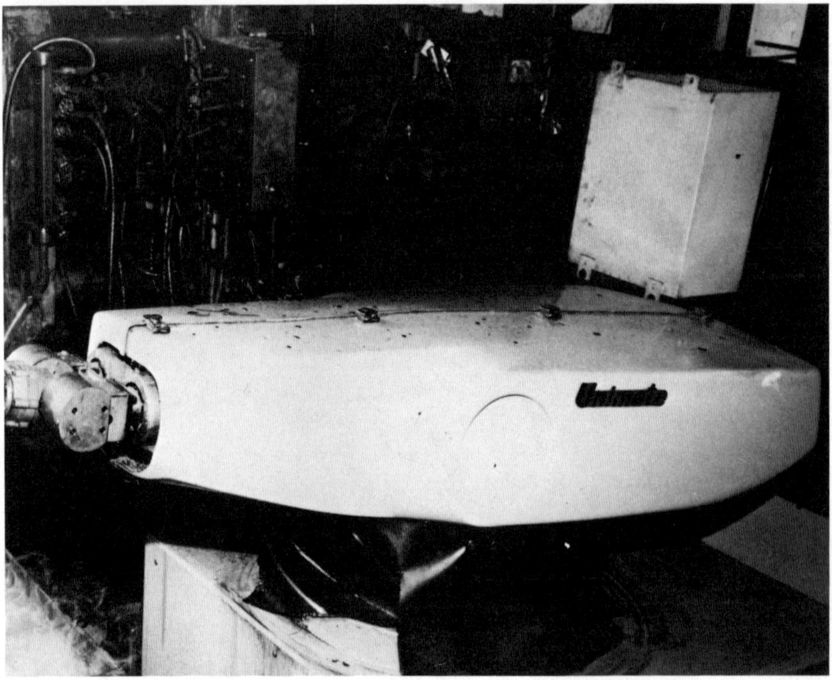

This Unimate robot is putting a burning hot object into a tank of cold water. A human hand could not touch anything that hot.

3
INDUSTRIAL ROBOTS

ROBOTS IN THE FACTORY

There's a new worker on the job. She can learn to do work fast. She doesn't make mistakes. She'll work all day and night. She'll never complain. She never will take a day off. She doesn't get sick. Besides all those things, she doesn't care about pay. This someone isn't a person. She's an industrial robot.

The industrial robot isn't a new idea. It has been in use since 1961. The Unimate robot was the first. It was invented by Joseph Engelberger. Unimation is the largest maker of robots in the world. Mr. Engelberger is the president of the firm that is now a part of the Westinghouse Corporation.

Companies in Japan saw a need for the robot. In 1967 they began to buy them. They bought the Unimate. They felt that the robot could do the job of many people. They were right. The robots were used to make cars.

Today Japan has most of the robots in the world. They have more than 12,000 robots. They work on the assembly line. On an assembly line an object moves on something like a moving shelf. People or robots are on both sides of it. As the shelf moves by, they do their job. Each worker does a part of the job. One person doesn't make the whole thing. One person does the same job all day. Most of the robots in Japan work in the car plants.

The Japanese are very happy with the work of their robots. They have opened a robot factory. In this factory each robot works for twenty-four hours a day. What does the robot do? It makes new robots. The plant will make one hundred robots each month.

The United States is also using a lot of robots. There are now more than 4000 robots at work in the U.S. About 2000 new robots are built here each year. Almost half of our robots are used in auto plants.

Robots have been around for almost thirty years. Companies have not used them much until now. Why? Is it a fad? Can't companies find enough people to work?

Robots have become popular for one reason. The reason is cost. Years ago the robot was expensive to make. It still costs a lot. It costs about the same or a little more than it did twenty years ago. Workers' pay has gone up much more. The workers are now earning five times as much as they did twenty years ago. They were earning about four dollars an hour then. Now at twenty dollars an hour companies have a hard time making a profit.

The industrial robot can do many jobs. It's not just a machine. It can stack plates one day. The next day it can paint. On the third day it can paint wagons. A simple machine does only one job every day. It can only do the same thing all the time. That's why an industrial robot is so valuable.

One company makes electric robots. They say that its robot can do many things. It can hang from the ceiling. It can move along a track. It can work in a room with things that may explode. It can grind big bits of metal and weld things together. It polishes sinks. It glues parts together. It can even inspect people's work. The robot can do it all without a human. The computer tells it what to do. The robot does the rest.

The industrial robot has a strong arm. It can do most of what a human arm can do. It learns its job well. A human leads it by the hand. The human leads it step by step. The steps are recorded in its memory. Then it can do the job perfectly. If the job changes, it can be taught something else. It only takes minutes to teach the robot a new job.

How is it all possible? How can one robot do so many different jobs? A holder is put on the end of the hand. The holder is like a pair of fingers. It's called a gripper. Grippers open and close. They can pick up tubes and plates. Then they can stack

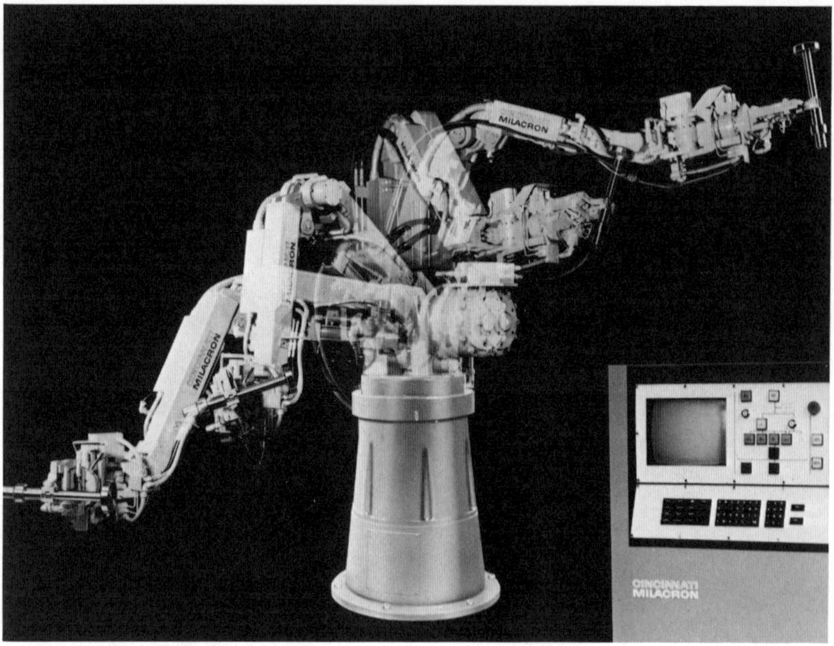

Humans must first lead this industrial robot step by step. Then the steps are recorded in the computer and the robot can do the job by itself.

them. Grippers can pick up things of many shapes. They can pick up small or large objects.

Grippers can do other jobs as well. Sometimes two gripper hands can be put on the same robot arm. Then the arm can do two jobs. One gripper hand can load things while the other gripper hand can unload something else.

The robot's arm may hold something besides grippers. The hand may hold tools. It can hold a spray gun to paint. It can hold a gun to weld. Believe it or not, it can cut glass and write with a pen. The robot is a very handy worker.

INDUSTRIAL ROBOTS ON THE JOB

The Unimate was the first industrial robot. The first one was bought by General Motors. In 1962 GM bought it. The robot was to be used to make cars. Its job was to weld car parts. The Unimate had a lot of problems then. It took many hours to set it up. It broke down often.

The Unimation Company worked hard to fix its robot. It's now the most widely used industrial robot in the world. Today Unimate robots can be set up with ease. There are six different kinds. Some have to be mounted on the floor. Others can be set up anywhere. They work well in hot or cold spots. They can work in a place as cold as forty degrees Fahrenheit and can work in a place as hot as one hundred twenty degrees Fahrenheit.

The Unimates are all strong. Some are stronger than others. The Unimate 1000 can lift fifty pounds. The Unimate 4000B can lift 450 pounds. The other four models can lift much more than the Unimate 1000, but less than the 4000B. The Unimate 2100C lifts up to 270 pounds. The 2000B, 2100B, and 2000C each can lift up to 300 pounds.

Imagine this. You are moving to a new house. The mover shows up with the Unimate 2100B and 2100C. The Unimates grab your chairs and bed with their grippers.

The Unimates have been able to move heavy things. They can hold heavy welding guns as well. There are quite a few companies that didn't need a robot to lift heavy things. Instead, they had light things to move. The things had to be moved quickly. There was a need for a smaller robot. It had to take up a small space. So the PUMA robot was invented.

The PUMA is a mini robot. It's the size of a human arm and hand. It works side by side with humans. It can handle any light materials. It moves just like the human body. It has a waist, a shoulder, an elbow, and a wrist. It can work at the same speed as humans. A boring job doesn't slow down the PUMA. A dangerous job doesn't scare it either. The PUMA keeps on working.

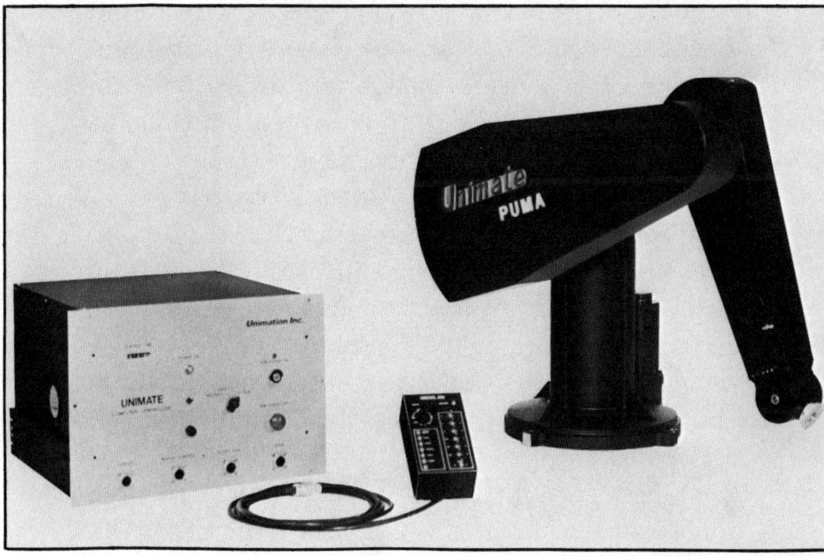

The PUMA is a mini robot that is the size of a human arm and hand.

The PUMA is used where small parts are put together. The PUMA helps assemble small appliances. It can work on toasters, irons, and can openers. It puts things into packages. PUMA puts things into wooden boxes and seals them. The PUMA may be small, but it does a big job.

Teaching the Unimates and PUMA isn't hard. It can be done two ways. The first way is to do it by hand. Move the arm to the points you want. Then record the points in the computer's memory. The second way is to use the computer terminal. Give the information to the terminal. The computer will set up its own program. It tells the order in which to do things. Then it tells the speed. Finally it tells the robot how it will move. You can leave the rest to the robot.

Another industrial robot is T^3. The T^3 is made by Cincinnati Milacron. It was the first firm to make a robot that worked on moving cars. Like the Unimates, the T^3 is easy to set up. It takes a few minutes. It can do a lot of the same things as the Unimate. The T^3 can load and unload tools and machines. It can move materials from station to station. It can pick up things and then package them. The T^3 does it all with its arm. The arm can move in six different positions. It can lift 225 pounds.

One of the most important things the T^3 does is weld. When a robot welds, it heats metal

One of the most important jobs the T^3 does is weld.

together. Welding is a dangerous job. With the T^3, no human must be there. No matter how much smoke, T^3 keeps working. The job is called arc welding. The torch moves along a straight line to join a seam. An operator watches the robot on television. It's much safer that way.

The T^3 helps make airplanes. It picks up airplane parts two at a time and takes them to a machine. Holes are drilled there. Then the parts are fastened together. The T^3 knows how to hold each part. It makes sure that the parts are held exactly the right way. The robot does not make a mistake.

Look at your refrigerator. You can thank a T^3 for it. The T^3 helps to make it. First T^3 puts a plastic liner into a press. Another T^3 is used to take it out. The liner is hard for humans to hold. The T^3 does it with ease.

GE brings good robots to life. General Electric is the name of the firm. Allegro is its new robot. It helps in the factory. It's used to assemble small things. It works best with tiny parts. The robot fits small parts inside car engines and air conditioners. Allegro puts things together. That's why it's called an assembly robot.

The Allegro robot is controlled by two computers. The robot has a sense of touch. It can tell if there are parts missing. It can also tell if parts are broken.

Allegro can work in a number of ways. Its arms

The Allegro is one of the most advanced of the industrial robots.

can work side by side with humans. Its arms can work side by side with other Allegro arms. It saves money. The quality of the products is high. Few things have to be done over. More products are made in less time.

The Allegro robot can be used to do many different jobs. It can pick things up. It can place screws where they are needed. Things can be fastened together by it. Things can be bent to the right shape or size. When the time comes to do a different job, Allegro is ready. It's felt that Allegro is the most advanced of the industrial robots.

Allegro, PUMA, Unimate, and T^3 all do hard tasks each day. Most of the more than five thousand robots in the U.S. lift things. Often it's a dangerous job. They can do the heavy work. They can handle chemicals or gases that could hurt us. There are dozens of other things that humans don't like doing. Robots take their place and do a great job.

ROBOTS THAT MAKE CARS AND TRUCKS

Of every ten industrial robots at work, four build cars. The uses for robots seem to grow all the time. Companies look to make a better car. They also hope to keep down their costs.

It happens in Detroit. Objects nine feet long move toward car bodies. They hiss like snakes. Sparks start to fly. It's the plant where the Chrysler

Robots are used to weld the sides of the Chrysler K-cars.

K-cars are being made. There used to be 200 people with welding guns on the job. Now they aren't there. In their place are fifty robots. The robots work two eight-hour shifts.

There are ten station auto-body welding lines. They're automatic. It takes less than two minutes to weld a car. The humans who had been welders are now free to work in a different part of the plant.

Each K-car gets almost 3000 welds. They call the system Robogate. In Robogate the two sides of a car and the underbody are welded together. The welding is done automatically, all by robots.

Robogate is smart. It can tell the styles of cars. It knows if a car has two doors or four doors. If there's a station wagon, it knows that, too. All welding is controlled by computer. A Master Control Panel controls everything. It checks the actions of the robots. In just one year, Robogate worked on almost 600,000 K-cars. That was in 1981.

Car companies try to find new ways to use robots. The Volvo plant in Sweden works with the Robogate, too. Twenty-seven of them are at work there. The robots freed seventy workers to do other jobs.

General Motors has found a new way. It's called Consight. It's a robot with eyes. The robot has a camera. The camera looks at parts as they move along. If they aren't in the right order they're pulled out.

The Robogate system is used to weld together the two sides and the underbody of the Chrysler K-cars.

Mack Truck has a plant in Maryland. Once humans built the Mack Truck, but not any longer. Nordson robots help. Engines need painting for protection. Humans had done the job for years. There were parts that were hard to reach. After doing the job, workers had to touch up missed spots. This took extra time. It also cost the company money. Nordson had an answer.

Nordson came down to Mack Truck. It trained the firm in the use of their robot. Now the Nordson robot does a perfect job. There's no need to touch up. The robot doesn't miss spots. Now it uses half as much paint as before.

General Motors is the largest car firm in the United States. It was the first one to buy industrial robots. Now, many years later, it has almost one-tenth of all the industrial robots in the U.S. at work. The first ones were used to weld parts together. Now the robots do many things. They spray-paint cars. They measure openings for doors and lights. They handle car parts on an assembly line. They still do welding. Now they can do the job in less than half the time. It used to take forty or fifty steps to weld by hand. Now it can be done in two or three steps.

Many people think the auto workers don't like the robots. That's not true. At a Ford plant in Chicago a robot broke down. The workers had a pet name for it. They had called it "Clyde the Claw."

The Ford workers gave Clyde a get-well party after it broke down.

Robots at car plants are here to stay. Sometimes they do a better job than humans. They may also do things much faster.

4
OFFICE ROBOTS

Don't look now. That worker bringing the coffee may not be a person. It could be a robot. Robots are now in the office, too. If you go to a metal company on Long Island, New York, you'll be in for a surprise. A robot greets you as you walk in. He blinks his eyes. He turns his head. Then he asks who you are. Actually, there are workers sitting in another room. They're watching you on television.

There is ESR-1. What a robot that is! It weighs 980 pounds. It has a head that turns around 360 degrees. Inside the head and body there are four television cameras. What is it? It's a robot security guard.

The ESR-1 can do more than sound an alarm. On a television show it took an egg out of John Davidson's hand. It opened a door. Then it pushed its way through the door. The ESR-1 can move in any direction. It obeys commands from up to a mile away.

Citibank has robots that are going places. They're robots on carts. The robots go up and down each hall. They drop off and pick up mail. There is an invisible chemical sprayed on the floor. The robots follow the path of the chemical. On the bottom of the cart there's a special light. The light finds the chemical as the chemical glows. The robot goes wherever it sees the chemical.

The robot is called Citipage. It moves slowly. It only goes one mile an hour. That's about the speed of someone walking. It has two blue lights that flash. It beeps to tell people that it's coming. The robot can't hurt anyone. As soon as it touches someone, it will stop. Its sensors stop it right away.

In West Germany, Charley is admired for his clothes. He always wears the latest styles. You might say he's a model. You might also say he's a robot. Charley works at the Institute for Clothing Science. When he wears clothes, people study him. On his skin, Charley has sensors. He acts the way a human acts. He may be comfortable. His shirt may be too tight. His pants may make him itch. The sensors tell people whether Charley is comfortable or not. Charley can even perspire. You can't get more real than that!

Nano doesn't perspire. He does move around a bit, though. Nano is a robot salesperson. He works at

The drink caddy robot has a supply of cold drinks and ice. It will serve you refreshments on its tray.

J&B

an office-supply store in Aurora, Colorado. Nano is a smart shopper. He knows everything that is on sale. Nano walks throughout the store. He points to things on sale. Nano tells the shoppers about the best buys. Shoppers enjoy Nano's tips.

5
ANDROIDS

"Come over here, you." You look up, but no one is there. All you see is a large cone. "What is your name?" Is this some kind of joke, you ask yourself. Actually, it's your first meeting with an android. An android is a robot that looks human. Or sometimes it can do the same things a human can.

The androids are Display Robots. Quasar Industries is their maker. They are used to tell people where to shop. They also entertain. When people are entertained, they remember the store and go back.

The Display Robots are sixty-four inches tall. That's five feet four inches. They're tall enough not to scare people. They're also the size to show they're not a toy or stuffed doll. They weigh 240 pounds. That's more than most people weigh. Still, they can travel anywhere. They can go in a car, plane, or ship. Inside the body there are many parts. There are eight-track tape decks and there are speakers. Batteries are inside, too. All these things help the robot act almost human.

It's hard to believe. The robot has a 4800-word vocabulary in English. It can be taught other languages, too. It has a humanoid voice. It works an eight-hour day.

Klatu is one of the Display Robots. He can work anywhere. It doesn't matter if he's indoors or out. It can be spring or fall, night or day. The robot is ready to do it all. Klatu likes people. He is often invited to meetings and stores. He likes to be a star. The robot brings helpers with him because he doesn't like to work alone. He likes to answer people's questions.

One time Klatu was an announcer at a meeting. There were twenty-two different speakers. Klatu introduced each one to the people. He did more than that. He made a joke about each person first. Everyone loved Klatu.

Klatu goes to business shows and to shopping malls, too. Newspaper and TV reporters ask him questions. He answers them all. He gives out free things. He cuts ribbons. Klatu talks with the people that pass by. Sometimes he appears on television talk shows. The robot is able to keep people happy.

Sentry II is another android. It's also made by Quasar Industries. Sentry II is not as friendly as Klatu. Sentry II is a guard robot. It can work for

Klatu is a display robot that is often invited to meetings and stores.

twelve hours. Then its battery must be charged. It's fast. It can move up to thirty miles an hour. Sentry II is eight feet tall. It weighs 650 pounds. It's bigger and heavier than a Display Robot, but its shape is the same.

If Sentry II spots an intruder, it shouts. It says, "Do not move." It also warns the intruder not to touch it. A two-way radio hears everything that happens. Sentry II can grab a person's arm. If it has to, it makes a very loud blast. The ear pain causes the intruder to let go of the robot. Then bright lights are used. If nothing works, Sentry II shoots laughing gas. The laughing gas will make the intruder feel sleepy. He will either run away or fall asleep.

Sentry II and the Display Robots both have hard jobs. Some people say they are something out of this world. They're from the space age and beyond.

Even harder to believe are Adam and Andrea Android. They, too, are android robots. Adam and Andrea are in human form. Adam is six feet tall. He has curly brown hair. He has a short beard and a moustache. People seem to love his smile. Andrea is five feet nine. She has a body like a movie star. She loves to disco.

At their feet is a Robot Cart. Inside are a microphone and tape deck. They are controlled by a box

Adam and Andrea Android are android robots that are controlled by a box someone holds.

someone holds. The androids can move up to three miles an hour. They shake hands. Their heads can turn. Their eyes are made of glass. That makes them look even more true to life. You can have Adam or Andrea if you pay $10,000!

6

FAMOUS ROBOTS ON FILM

The most well-known robot was the Frankenstein monster. In the film a scientist made a human robot. It followed the scientist's orders. It could do a lot of things.

Most robots that followed Frankenstein in films didn't look human. In *Earth Vs. The Flying Saucers* the robots didn't look real. They seemed too mechanical. They had two arms and two legs but they were quite stiff as they moved.

A lot of robots were evil. They wanted to destroy. In *Kronos* there was a giant robot. He was more than one hundred feet tall. He came from another planet to take Earth's power. The Earth's power was at last used to destroy Kronos.

Another monster robot was Gort. Gort was in *The Day The Earth Stood Still.* He was ten feet high. He had wide shoulders. Gort and his master, Klaatu, came to warn Earth. They told Earth not to use nuclear power. Gort also talked.

The robot in *Colossus of New York* was larger. This one was twelve feet tall. A doctor put the brain of his son into a robot. It began to attack people. It looked like a statue in Central Park.

Target Earth had many monster robots instead of one giant one. In the movie, all that is left on the Earth is one man. The rest are robots from space.

Not all robots in the movies were bad. Robby the robot was good. Robby was in *Forbidden Planet.* He may have been the smartest robot ever. He knew one hundred eighty-seven languages. Robby lived with a few people on the planet Altair IV.

And then there were R2D2 and C3PO. They were good robots, too. They were in the movie *Star Wars.* R2D2 had a tiny man inside. The man was Kenny Baker. He sat inside the robot. His arms were pressed at his side. A lot of people thought the robots were the best part of *Star Wars.*

Sometimes there have been good and bad robots in the same film. The show *Logan's Run* had both. Rem, the android, was a good robot. He looked just like a human. On the same show there were robot guards. They protected a strange woman from everyone. The robots were very strong.

In *The Black Hole* there were also robot guards. They looked like metal men and they carried guns. The robots had a battle with the people from another spaceship. There was also a killer robot in the movie. His name was Maxamillian.

Rem, the android, was a good robot who looked just like a human on the show Logan's Run.

Westworld had robots that tried to help. They were androids. They looked like you and me. The robots were there to make visitors happy. Each person played out secret dreams. Then the robots stopped following orders. They tried to take over Westworld. In the end, people had to take apart the robots.

Robots in movies have always been fun to watch. They were make-believe. People said that there couldn't be real robots. Now there are. Robots can do more and more exciting things.

7

ROBOTS OF THE FUTURE

Robots are getting smarter all the time. They'll soon be able to do even more amazing things. Robots are now being made to see and touch. They can understand what's in the camera. They know the size, shape, temperature, and softness of what they touch.

The Navy is working on a new robot. This kind will ride alone in a submarine. It will find ships that have sunk. Then it will fix them.

At Waseda University in Japan, they're working on a robot. It will be able to see and hear. It will also be able to understand what you tell it. It will be as smart as a five-year-old. It will be able to watch someone build with toy blocks. Then it will build the same thing.

A firm in Japan is trying to help fishermen. They're building a robot. It will separate fish. It will tell one kind of fish from another. It will also put fish together by their size.

Quasar Industries already has Display Robots. Now it's working on a home robot. It will be ready in a few years. The robot will serve drinks. It will clean the carpet. The robot will answer the door, too.

General Motors will have a robot car by the year 2001. You'll push a button. Then you'll say, "I want to go to work." The car will take you to work. You'll also be able to park at the shopping mall. Call the car with a radio transmitter. The car will pick you up.

At Stanford University they have an industrial robot arm. It's used by people who can't use their arms and legs. The person's voice tells the arm how to move. Some day all such handicapped people may have this robot.

At the Veteran's Hospital in Palo Alto there's a different robot arm. It can do even more things. It can pick up a phone. It can hold a book. It can hold a glass. The arm can even pour a drink. This arm is now being tested.

Not too far from now there may be more home robots. There will be a washing machine with an arm to unload the clothes. There will be a lawn mower to do the job by itself.

Here are some predictions of the future. The University of Michigan predicts these may happen:

- By 1985, one-fifth of the work at car plants will be done by robots.
- By 1988, one-half of all small parts in factories will be put together by robots.

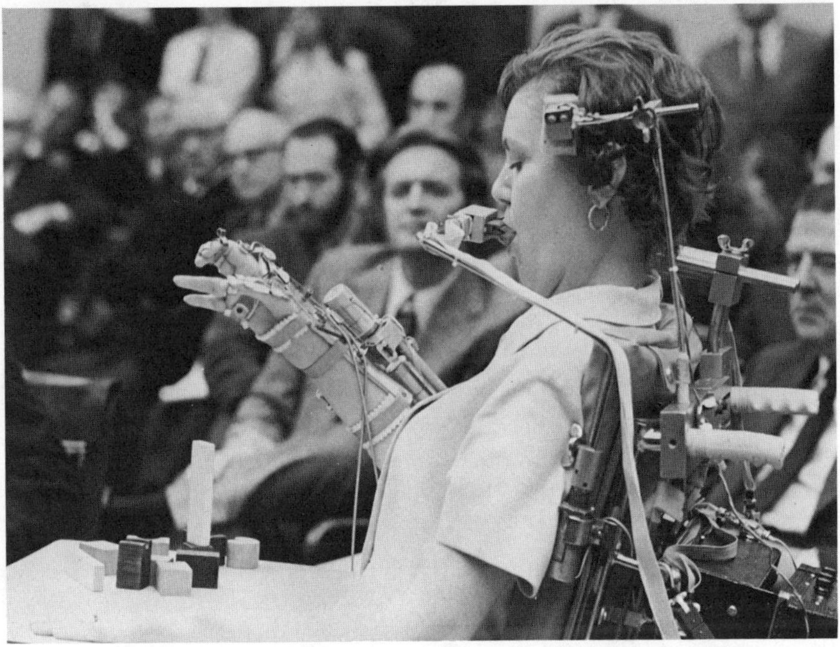

When this woman speaks into the microphone she will be able to move her arm.

- By 1990, robots will be able to do almost all of the work that calls for putting things together.

Computer scientist Marvin Minsky has a prediction. In about twenty years he believes that operations will be done in a different way. Doctors will put on special gloves. They will be joined to mechanical hands. These special hands will do the operation.

Someone else has done work on a new kind of robot. It can be used in a hospital. It will lift up the person from the bed. Then the nurse will change the sheets. The robot will put the person down and the nurse will then cover the patient.

James Clark of Westinghouse has thought about robots. "If a robot can do the job, a person should not be doing it anyway." In the future, workers will be doing many new jobs.

Robots won't take over. They'll be working with us side by side. Robots are here to stay.

Would you like to play a video game with this robot of the future who is here today?

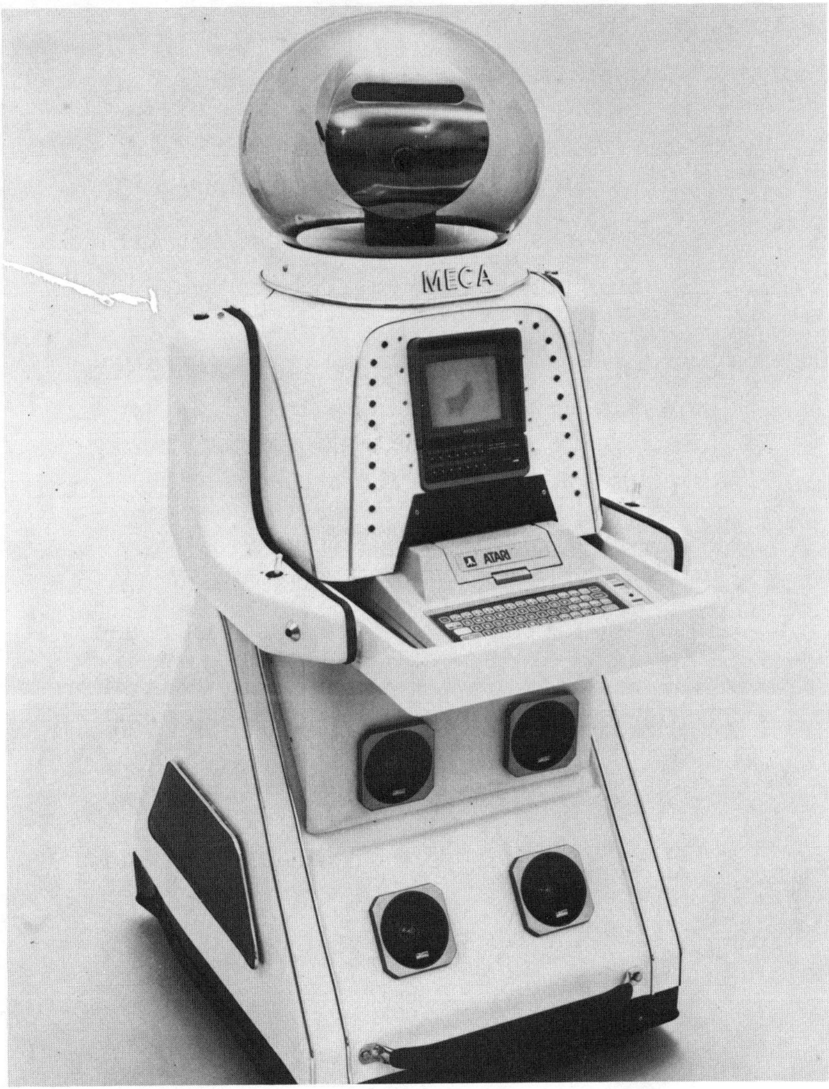
MECA
ATARI

GLOSSARY

Android—a robot that looks human and/or sometimes can do the same things a human can.

Computer—an automatic electronic machine that is often used to program robots to do different jobs.

Fahrenheit—a scale to measure temperature on which the boiling point of water is 212 degrees above the zero of the scale and the freezing point is at 32 degrees above zero.

Feedback—the return of information to the beginning source of the information.

Gripper—a holder put on the end of a robot's hand that works like a pair of fingers.

KRM or Kurzie—Kurzweil Reading Machine. This robot looks like a photocopying machine, but it can read the books that are in libraries.

Robot—a word in the Czech language that means worker.

R.U.R.—the name of a Czech play that introduced the word robot. *R.U.R.* stood for Rossum's Universal Robots.

FURTHER READING

Heiserman, David. *How to Build Your Own Self-Programming Robot.* Tab Books, 1979.

Rutland, Jonathan. *World of Robots.* Warwick Press, 1979.

Pienkowski, Jan. *Robot.* Delacorte Press, 1981.

Readers may also request additional information or direct questions to:

Android Amusement Corp.
2650 Myrtle Avenue B-7
Monrovia, California 91016

Quasar Industries, Inc.
The Robot People
59 Meadow Road
Rutherford, New Jersey 07070

Cincinnati Milacron
4701 Marburg Avenue
Cincinnati, Ohio 45209

Unimation Inc.
Shelter Rock Lane
Danbury, Connecticut 06810

INDEX

ABOUT THE AUTHORS

Stuart and Donna Paltrowitz are a writing team who could have used a robot to assist with their varied projects. They have had more than thirty books published for children and adults. Their articles have appeared in major magazines and newspapers throughout the world. Together they have created microcomputer software programs that have been published for home use. These native New Yorkers work at their Long Island home where they reside with their sons, Adam and Darren, and a cat, Marshmallow. Stuart is a graduate of Hofstra University and has a Master's Degree from Long Island University. Donna received her Bachelor of Arts and Master's Degree from Hofstra University.